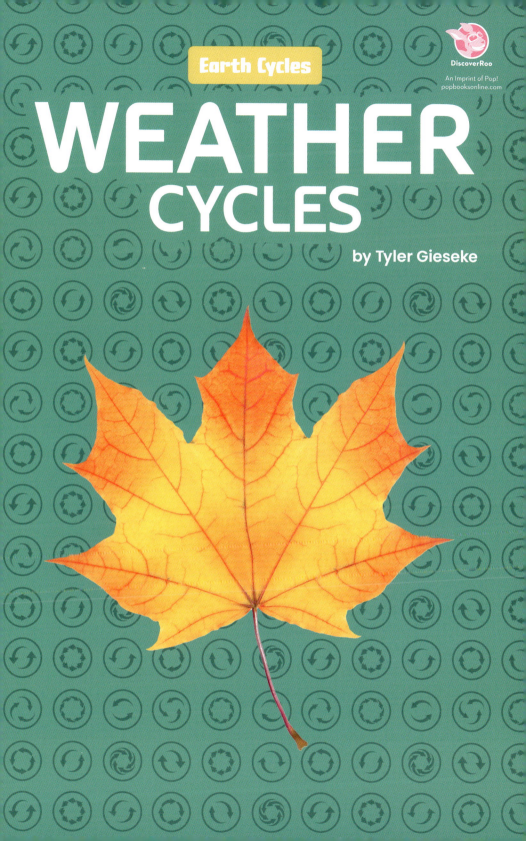

abdobooks.com

Published by Pop!, a division of ABDO, PO Box 398166, Minneapolis, Minnesota 55439. Copyright © 2023 by Abdo Consulting Group, Inc. International copyrights reserved in all countries. No part of this book may be reproduced in any form without written permission from the publisher. DiscoverRoo™ is a trademark and logo of Pop!.

Printed in the United States of America, North Mankato, Minnesota.

052022
092022

THIS BOOK CONTAINS RECYCLED MATERIALS

Cover Photo: Shutterstock Images
Interior Photos: Shutterstock Images; NASA, 20–21
Editor: Elizabeth Andrews
Series Designer: Laura Graphenteen

Library of Congress Control Number: 2021951861
Publisher's Cataloging-in-Publication Data
Names: Gieseke, Tyler, author.
Title: Weather cycles / by Tyler Gieseke
Description: Minneapolis, Minnesota : Pop, 2023 | Series: Earth cycles | Includes online resources and index
Identifiers: ISBN 9781098242244 (lib. bdg.) | ISBN 9781098242947 (ebook)
Subjects: LCSH: Hydrometeorological cycles--Juvenile literature. | Meteorology--Juvenile literature. | Weather--Juvenile literature. | Weather fronts--Juvenile literature. | Earth sciences--Juvenile literature. | Environmental sciences--Juvenile literature.
Classification: DDC 551.6--dc23

Pop open this book and you'll find QR codes loaded with information, so you can learn even more!

Scan this code* and others like it while you read, or visit the website below to make this book pop!

popbooksonline.com/weather

*Scanning QR codes requires a web-enabled smart device with a QR code reader app and a camera.

TABLE OF CONTENTS

CHAPTER 1
Change of Season................ 4

CHAPTER 2
Four Seasons..................... 10

CHAPTER 3
Dry and Rainy 16

CHAPTER 4
Let It Snow....................... 22

Making Connections.............. 30
Glossary 31
Index............................. 32
Online Resources................. 32

CHAPTER 1
CHANGE OF SEASON

It's late February in Minnesota. Piles of snow cover the cold ground. People wear thick coats and put on hats and gloves before they go outside. Freezing weather is normal for this part of the year.

WATCH A VIDEO HERE!

Ice-skating is a popular winter activity.

In just a few weeks, it will be spring. The temperature will rise, and the snow will melt. People will put away their

Springtime is known for pretty flowers.

DID YOU KNOW? Northern Minnesota has an average of 70 inches (178cm) of snow per year.

hats and gloves. They will bask in the warm weather for many months. But the snow and cold will be back again the next winter.

Seasonal changes like these are examples of weather cycles. Weather cycles are important Earth cycles that affect how hot or cold a place is and how much rain or sun there is. Weather cycles often repeat every year.

Not all weather cycles are the same. Some **regions** of the world have four seasons: spring, summer, fall, and winter. Others have only two seasons: a dry season and a rainy season. In the **Arctic** and **Antarctic**, the weather is cold and snowy all year round.

People and animals come to expect the weather to happen a certain way in each area. Understanding weather cycles is important!

SEASONAL CYCLES

FOUR SEASONS

Spring
- warm
- stormy

Summer
- hot
- sunny

Winter
- cold
- snowy

Fall
- cool
- leaves drop

TWO SEASONS

Rainy Season
- heavy rain
- rivers fill

Dry Season
- little rain
- rivers may dry up

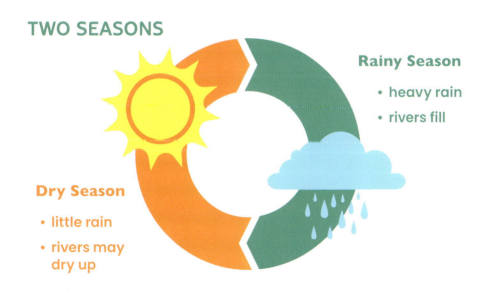

CHAPTER 2
FOUR SEASONS

Temperate parts of the world often have four seasons. In the summer, the air is hot. The weather is usually sunny or rainy. It starts to cool down in the fall. Leaves change their color and drop from

LEARN MORE HERE!

The four seasons are very different.

the trees. The winter is cold and usually snowy. Finally, the air starts to warm and the snow melts in the spring.

EARTH'S SEASONS

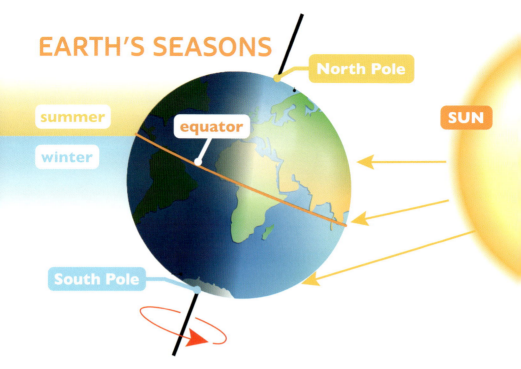

In this image, Earth's axis is the tilted black line.

This weather cycle is most common in places halfway between the **equator** and the North or South Pole. This includes central North America, Eastern Europe, and the southern **regions** of South America and Australia.

The four seasons happen because Earth's **axis** is tilted. For half of the year, the northern part of Earth is pointed toward the Sun. It is spring and summer there. The southern half is pointed away from the Sun. So, it is fall and winter there.

Six months later, Earth is on the opposite side of the Sun. But its axis still tilts the same way. Then, it is summer in the south and winter in the north.

DID YOU KNOW? Earth's axis is tilted by about 23.5 degrees.

FORMING TORNADOES

Tornadoes usually form from a thunderstorm called a supercell. This means there is warm, moist air twisting upward at the center of the storm. The twisting air is called a vortex.

The vortex begins to tilt as the storm grows. This pushes cold and dry air to the ground. The moisture in the warm vortex turns it into a funnel cloud. The cold air forces the funnel cloud into a small space. If they are strong enough, the hot and cold air push the funnel cloud to the ground. It's a tornado!

Temperate regions experience thunderstorms and tornadoes, particularly during spring. Cold air sits above the North and South Poles, and the air above the equator is warm. The cold and warm air meet in temperate regions.

Tornadoes can destroy most things in their paths.

When seasons change, the warm and cold air mix. This can create storms.

CHAPTER 3
DRY AND RAINY

Tropical parts of the world can have just two seasons. The temperature remains about the same throughout the year. But the amount of rain that falls and water that is in the ground changes. During the dry season, there is little rain. People,

EXPLORE LINKS HERE!

plants, and animals must be careful not to run out of water. In the rainy season, the rain comes much more often.

Plants grow more easily in the rainy season.

This weather cycle is most common in places near the **equator**. These include Central America, Brazil, central

The equator runs through Ecuador, Colombia, and Brazil in South America.

DID YOU KNOW? The African lungfish can survive a dry season by burying itself in mud. It can live without water!

Africa, southern India, and Southeast Asia. Northern parts of Australia can also experience dry and rainy seasons.

Tropical **regions** get the most direct sunlight on Earth. So, their temperatures stay warm or hot for the entire year. But the tilt of the Earth's **axis** spreads out rain across the seasons.

When the Sun shines on the oceans near the equator, the water is heated, and some of it rises as **water vapor**. When the water vapor is high enough, it cools and forms clouds. Soon, the water falls back down as rain. The line where

The line of clouds shows the tropical rain belt.

this occurs is called the tropical rain belt. The tropical rain belt is north of the equator when it is summer there. The belt shifts south of the equator later in the year.

CHAPTER 4
LET IT SNOW

Polar areas are cold and snowy most of the year. They are far from the **equator** and do not get direct sunlight. But their temperatures are still warmer in the summer and colder in the winter.

COMPLETE AN ACTIVITY HERE!

Large drifts of snow and ice lie near the North Pole.

DID YOU KNOW?

The **Arctic** reaches about –35°F (–37°C) in the winter and as much as 45°F (7°C) in the summer.

Polar bears live in Arctic regions.

The amount of sunlight in polar areas shifts with the seasons. When it is summer in the Arctic, the sun can be out for the entire day. At the same time, it is winter in the **Antarctic**, and the sun might never rise. Six months later, the opposite is true.

Animals in the polar **regions** must find ways to deal with several weeks without sunlight. For example, reindeer can see better in the dark than other animals. Wolves and polar bears have an excellent sense of smell to help them find food in the dark.

Reindeer can see kinds of light that humans can't.

Weather cycles are important Earth cycles. Changes in temperature, rain, wind, and sunlight affect plants and animals and how they live. Weather cycles make Earth a very interesting place.

A flash of lightning can be beautiful.

Weather cycles are different in different parts of the world. This diagram shows the weather when it is summer in the Southern Hemisphere. That means it is also the rainy season just south of the **equator**. What would it be like in the Northern Hemisphere?

MAKING CONNECTIONS

TEXT-TO-SELF

Would you prefer to live in a tropical, temperate, or polar region? Why?

TEXT-TO-TEXT

Have you read other books about weather or climate? What did you learn from those books that you didn't in this one?

TEXT-TO-WORLD

What weather cycles happen where you live? Can you explain why they happen?

GLOSSARY

Antarctic — the cold, southern section of Earth.

Arctic — the cold, northern section of Earth.

axis — an imaginary tilted line through Earth that connects the North and South Poles.

equator — an imaginary line around the middle of Earth.

polar — having to do with the Arctic or the Antarctic.

region — a large area.

seasonal — having to do with seasons.

temperate — not too hot and not too cold.

tropical - having warm weather for most or all of the year.

water vapor — the gas form of water.

INDEX

Africa, 18–19
Asia, 19
Australia, 12, 19

Europe, 12

India, 19, 29

North America, 12, 18
North Pole, 12, 14

South America, 12, 18, 28
South Pole, 12, 14

ONLINE RESOURCES
popbooksonline.com

Scan this code* and others like it while you read, or visit the website below to make this book pop!

popbooksonline.com/weather

*Scanning QR codes requires a web-enabled smart device with a QR code reader app and a camera.